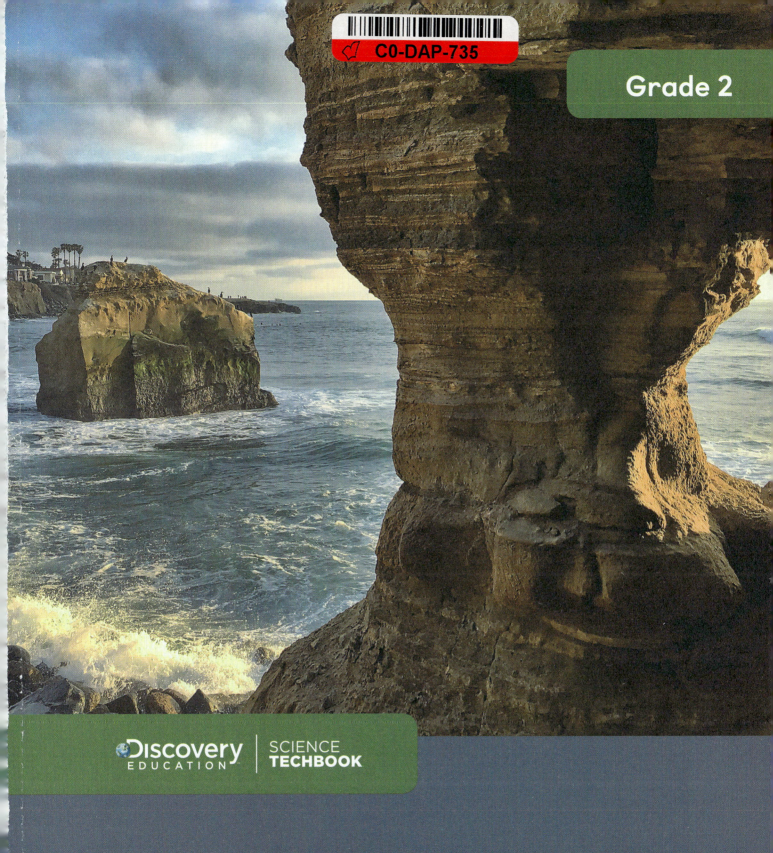

Grade 2

DISCOVERY EDUCATION | SCIENCE **TECHBOOK**

California
Unit 3
How Landscapes Change

To obtain permission(s) or for inquiries, submit a request to:
Discovery Education, Inc.
4350 Congress Street, Suite 700
Charlotte, NC 28209
800-323-9084
Education_Info@DiscoveryEd.com

ISBN 13: 978-1-68220-539-6

Printed in the United States of America.

3 4 5 6 7 8 9 10 WEB 23 22 21 20   B

Acknowledgments

Acknowledgment is given to photographers, artists, and agents for permission to feature their copyrighted material.

Cover and inside cover art: Sam Antonio Photography / Moment Open / Getty Images

© Discovery Education | www.discoveryeducation.com

Table of Contents

Unit Project

Grade 2 Resources

Discovery
EDUCATION

Dear Parent/Guardian,

This year, your student will be using Science Techbook™, a comprehensive science program developed by the educators and designers at Discovery Education and written to the California Next Generation Science Standards (NGSS). The California NGSS expect students to act and think like scientists and engineers, to ask questions about the world around them, and to solve real-world problems through the application of critical thinking across the domains of science (Life Science, Earth and Space Science, Physical Science).

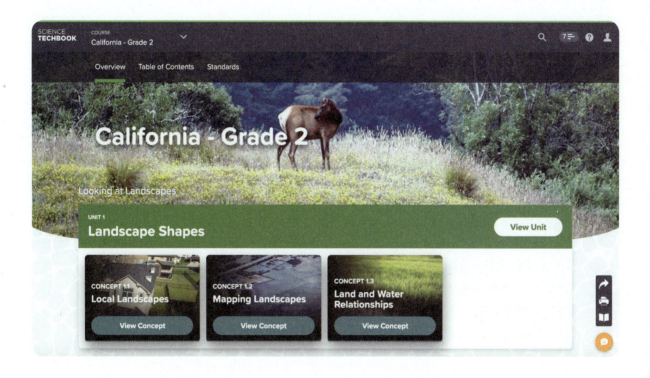

Science Techbook is an innovative program that helps your student master key scientific concepts. Students engage with interactive science materials to analyze and interpret data, think critically, solve problems, and make connections across science disciplines. Science Techbook includes dynamic content, videos, digital tools, Hands-On Activities and labs, and gamelike activities that inspire and motivate scientific learning and curiosity.

You and your child can access the resource by signing in to www.discoveryeducation.com. You can view your child's progress in the course by selecting the Assignment button.

Science Techbook is divided into units, and each unit is divided into concepts. Each concept has three sections: Wonder, Learn, and Share.

Units and Concepts Students begin to consider the connections across fields of science to understand, analyze, and describe real-world phenomena.

Wonder Students activate their prior knowledge of a concept's essential ideas and begin making connections to a real-world phenomenon and the **Can You Explain?** question.

Learn Students dive deeper into how real-world science phenomenon works through critical reading of the Core Interactive Text. Students also build their learning through Hands-On Activities and interactives focused on the learning goals.

Share Students share their learning with their teacher and classmates using evidence they have gathered and analyzed during Learn. Students connect their learning with STEM careers and problem-solving skills.

Discovery
EDUCATION

Within this Student Edition, you'll find QR codes and quick codes that take you and your student to a corresponding section of Science Techbook online. To use the QR codes, you'll need to download a free QR reader. Readers are available for phones, tablets, laptops, desktops, and other devices. Most use the device's camera, but there are some that scan documents that are on your screen.

For resources in California Science Techbook, you'll need to sign in with your student's username and password the first time you access a QR code. After that, you won't need to sign in again, unless you log out or remain inactive for too long.

We encourage you to support your student in using the print and online interactive materials in Science Techbook on any device. Together, may you and your student enjoy a fantastic year of science!

Sincerely,

The Discovery Education Science Team

Unit 3
How Landscapes Change

How the Land Changes Shape

The shape of Earth's surface can change. Some changes happen quickly. Some changes happen slowly. What causes the changes in the shape of Earth? If we know how Earth's surface changes, we can make predictions. We can also design solutions to slow down changes in Earth's surfaces.

How can you prepare for landscape changes? Watch a video about how one school prepares for a volcanic eruption.

Quick Code: ca2505s

Video

How the Land Changes Shape

Discovery
EDUCATION

Think About It

Look at the photograph. **Think** about the following questions:

- What evidence do natural processes leave behind as they shape Earth?

- How do the material properties of rocks affect what happens to them in landscapes?

Landslide Blocks a Road

Design Solutions Like a Scientist

Quick Code:
ca2506s

Hands-On Engineering: Landscape Changes

In this activity, you will design a solution to reduce or prevent landscape erosion.

Crumbling Cliff

| **SEP** Analyzing and Interpreting Data | **CCC** Structure and Function |

Discovery EDUCATION

Ask Questions About the Problem

You are going to design a solution to reduce or prevent erosion. **Write** some questions you can ask to learn more about the problem. **Write** down the answers to the questions.

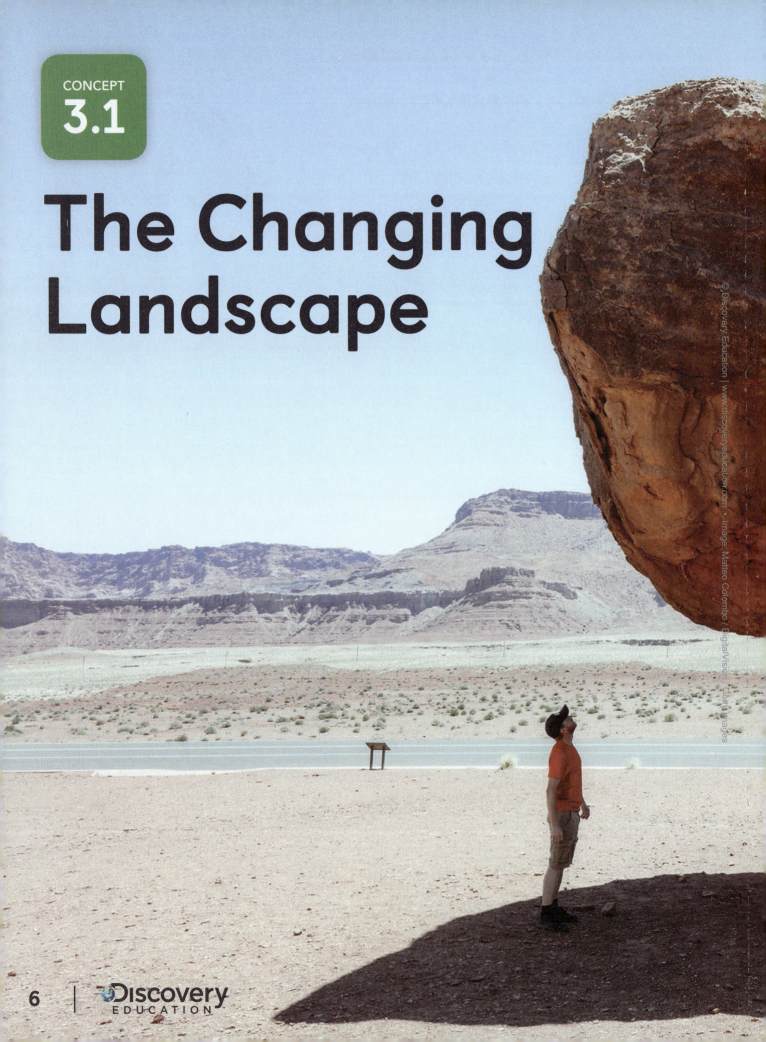

The Changing Landscape

Student Objectives

By the end of this lesson:

- ☐ I can observe patterns to explain how wind or water may have made the landform shapes.

- ☐ I can show evidence of landscape changes that happen fast.

- ☐ I can show evidence of landscape changes that happen slowly.

Key Vocabulary

- ☐ Earth's crust
- ☐ earthquake
- ☐ erosion
- ☐ weathering

Quick Code:
ca2508s

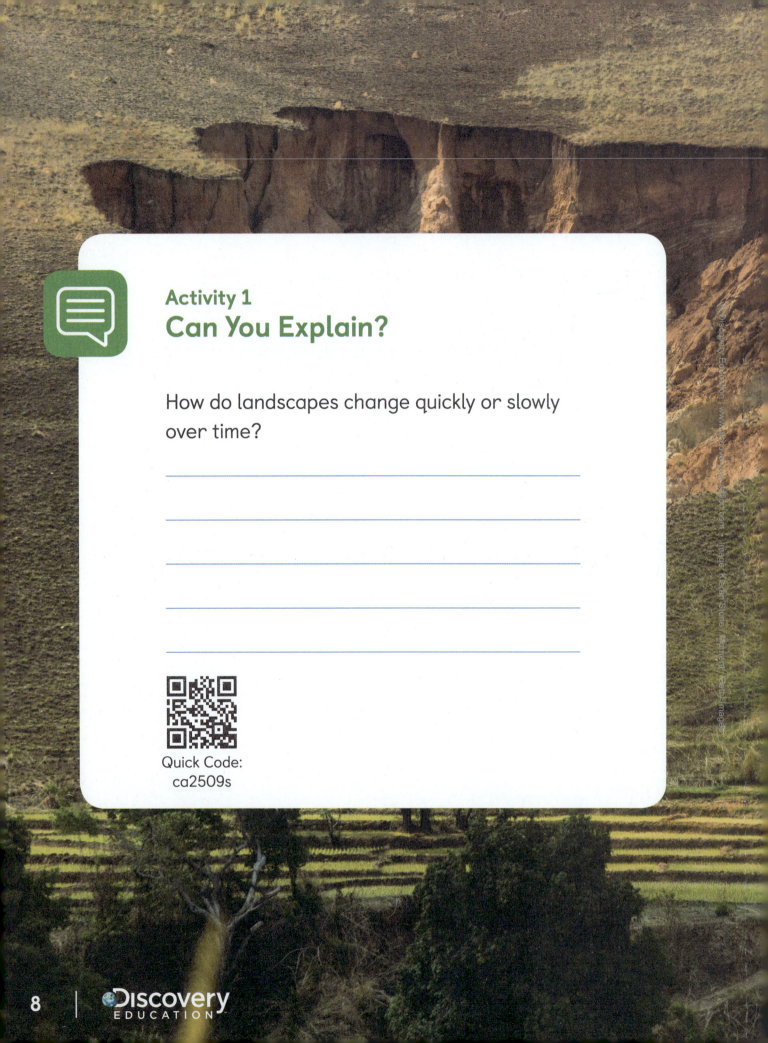

Activity 1
Can You Explain?

How do landscapes change quickly or slowly over time?

_____

_____

_____

_____

_____

Quick Code:
ca2509s

Discovery
EDUCATION

Activity 2
Ask Questions Like a Scientist

Quick Code:
ca2510s

House Mystery

Let's Investigate a House Mystery

Talk Together

There is a story behind every picture.

- What do you notice?

- Did this landscape change slowly or quickly?

- What questions would you ask to find out what happened here?

What do you wonder?

Activity 3
Observe Like a Scientist

Quick Code:
ca2511s

Mudslides

Watch the video. **Look** for the way mudslides happen.
Pay attention to how mudslides move.

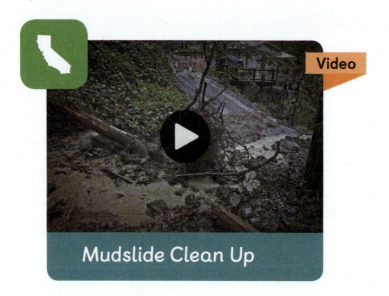

Mudslide Clean Up

Talk Together

Now talk together about mudslides. Talk about whether they
move fast or slow.

© Discovery Education | www.discoveryeducation.com • Image: Matteo Colombo / DigitalVision / Getty Images. Justin Sullivan / Staff / Getty Images. Icon: Freepik from www.flaticon.com

Activity 4
Observe Like a Scientist

Quick Code:
ca2512s

Cliffs and Mushroom Rock

Use your observation skills: **Look** at more landscape shapes. What do you see? What might have caused the shapes you see?

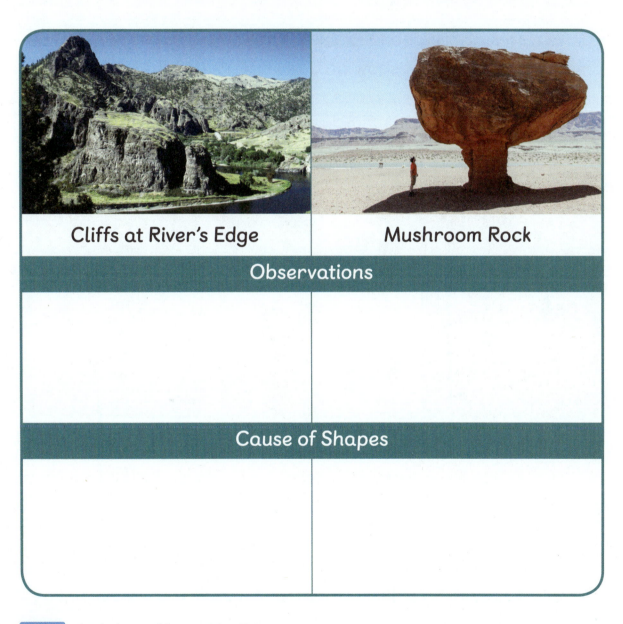

Cliffs at River's Edge	Mushroom Rock
Observations	
Cause of Shapes	

SEP Analyzing and Interpreting Data

© Discovery Education | www.discoveryeducation.com • Image: Matteo Colombo / DigitalVision / Getty Images (a) Paul Fuqua

Activity 5
Evaluate Like a Scientist

Quick Code:
ca2514s

What Do You Already Know About Changing Landscapes?

How Does Earth Change?

Rain is water. It can change things. **Look** at each image.
Circle the images that show an area before it rains.

Explaining Weathering and Erosion

Look at the image, and **answer** the question.

Winding River

How do you think wind and rainfall changed Earth and made this canyon?

_____

_____

_____

_____

_____

How Do Water and Wind Change a Land's Shape?

Activity 6
Observe Like a Scientist

Mitten Butte and Local Landscapes

When you look across your local landscape, what do you see?

Look at each image.

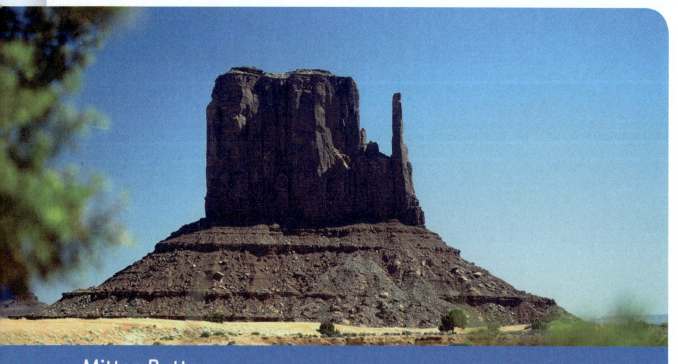

Mitten Butte

CCC Stability and Change

DISCOVERY EDUCATION

Rocky Mountains

Do you see hills, valleys, cliffs, or other interesting landforms?

_____

_____

How were these landforms made?

_____

_____

_____

What forces shaped the landscape?

_____

_____

Activity 7
Observe Like a Scientist

Quick Code:
ca2516s

Erosion

Watch the video and look for examples of weathering and erosion.

Video

Paul and His Science Pals

 Talk Together

What is the difference between ==weathering== and ==erosion==?

© Discovery Education | www.discoveryeducation.com • Image: SchoolMedia, Inc.; Paul Fuqua. Icon: Freepik from www.flaticon.com

Activity 8
Analyze Like a Scientist

Quick Code:
ca2559s

Weathering

Look at the picture. **Circle** the sentence in the paragraph that describes the picture.

Read Together

Weathering

Weathering is the breaking down of rocks into smaller pieces. Water can cause weathering. Sometimes, water can freeze inside a crack in a rock. The frozen water expands and makes the crack larger. The rock breaks apart into smaller pieces. Water flowing over rocks can make sharp edges smooth over time.

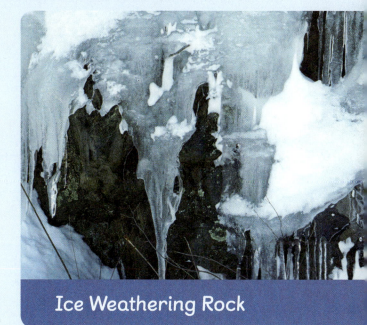

Ice Weathering Rock

Activity 9
Investigate Like a Scientist

Quick Code:
ca2517s

Hands-On Investigation:
Let It Rain

In this activity, you will look at how rain and wind can affect erosion and how plants change the process of erosion.

Make a Prediction

You are going to make a landform using sand. **Write** or **draw** your predictions.

How do you predict the wind will affect the landform?

How do you predict the rainfall will affect the landform?

What materials do you need? (per group)

- Aluminum foil pan, 13 in. × 9 in. × 2 in.
- Sand
- Watering can
- Salt shaker
- Spray bottle
- Water
- Disposable gloves (per student)
- Safety goggles (per student)
- Fan
- Artificial flowers or foliage
- Plastic pan
- Paper towels

What Will You Do?

Make your landform using the materials from your teacher. **Use** the fan to blow wind over your landform. **Write** what happened.

_____

_____

_____

_____

SEP Planning and Carrying Out Investigations

CCC Cause and Effect

Rebuild your landform. Use the salt shaker to make it rain over your landform. What happened?

_____

_____

Rebuild your landform. Use the spray bottle to make it lightly rain over your landform. What happened?

_____

_____

Rebuild your landform. Use the watering can to make it rain over your landform. What happened?

_____

_____

Rebuild your landform. Place a plant in your sand.
Try to make the wind and water change your landform again.
Be sure to use the same steps you used before. Why is that important? What happened?

_____

_____

_____

_____

Think About the Activity

How are wind and water erosion alike?

How are wind and water erosion different?

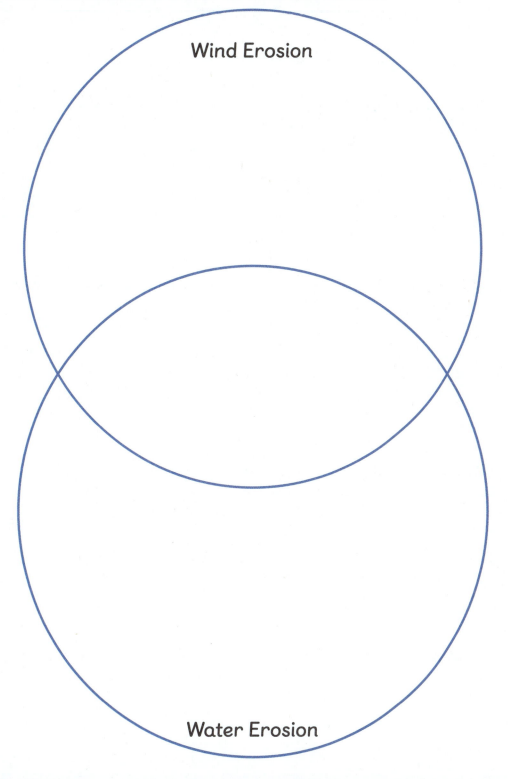

Wind Erosion

Water Erosion

How did your models show the ways rain and wind can change the shape of a landform? How was the change different with and without plants?

_____

_____

_____

_____

DISCOVERY EDUCATION

Activity 10
Evaluate Like a Scientist

Quick Code:
ca2518s

Wind or Water Erosion

Look at the pictures of landforms. **Write** what caused the shape of each landform.

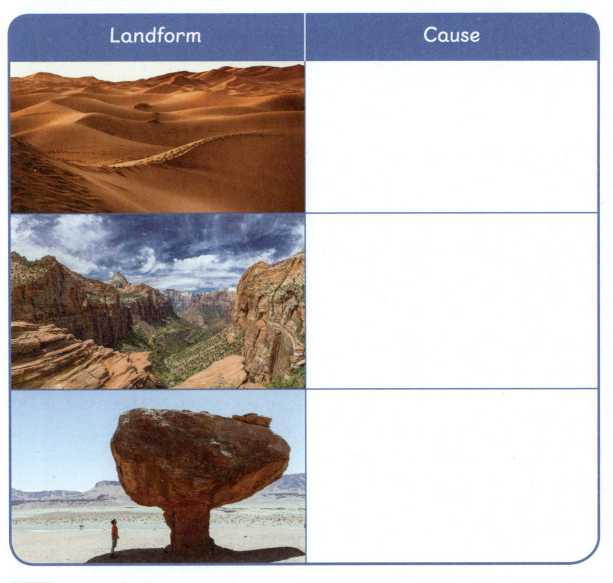

Landform	Cause

CCC   Cause and Effect

Slowly over Time

Rivers usually change the land slowly. Over time, a fast-moving river can erode rock and soil. The Colorado River has cut out parts of the Grand Canyon over thousands of years. Some rocks are harder than others. The hardest rocks can take a very long time to weather. Many landforms show evidence that softer rocks have eroded away while the harder rocks remain.

Colorado River Canyon

© Discovery Education | www.discoveryeducation.com • Image: Corbis Documentary / Getty Images

How Does a Land's Shape Change Slowly over Time?

Activity 11
Analyze Like a Scientist

Quick Code:
ca2519s

Slowly over Time

Think about what you have read. **Look** at the picture. **Answer** the question.

What other things do you know that change slowly over time? **Write** or **draw** your answer.

Activity 12
Observe Like a Scientist

Quick Code:
ca2561s

Shaping the Surface

Watch the video and look for evidence of weathering and erosion that happens slowly.

Video

Shaping the Surface

 Talk Together

Now, talk together about weathering and erosion. How do weathering and erosion shape the surface of Earth?

SEP   Obtaining, Evaluating, and Communicating Information

© Discovery Education | www.discoveryeducation.com • Image: Pixabay, Matteo Colombo / DigitalVision / Getty Images, Icon: Freepik from www.flaticon.com

Activity 13
Observe Like a Scientist

Quick Code:
ca2562s

Erosion and Deposition

Explore erosion and deposition online and **answer** the question.

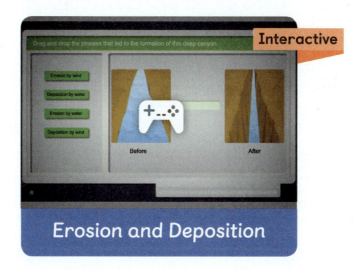

Erosion and Deposition

How can a computer model help us study something that happens slowly?

_____

_____

_____

_____

_____

SEP   Engaging in Argument from Evidence

Activity 14
Investigate Like a Scientist

Quick Code:
ca2563s

Hands-On Investigation: Rock Weathering

In this activity, you will predict and test the strength of two rocks when they are weathered by water. You will determine which type of rock weathers faster.

Make a Prediction

Do you think the chalk or the pebble will weather faster?
Write or **draw** your prediction.

SEP Planning and Carrying Out Investigations

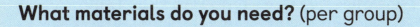

What materials do you need? (per group)

- Water
- Bowls, plastic
- Natural chalk
- Large pebble
- Eyedropper

What Will You Do?

Place the chalk in one bowl. **Place** the pebble in another bowl. **Show** with pictures or words how you will use the eyedropper to place water on the chalk and the pebble.

Use the eyedropper to place water on the chalk.
What happened?

_____

_____

_____

_____

_____

Use the eyedropper to place water on the pebble.
What happened?

_____

_____

_____

_____

_____

Which rock broke down faster and why?

_____

_____

_____

_____

_____

Think About the Activity

Draw what would have happened if even more water was dropped on both rocks.

Chalk

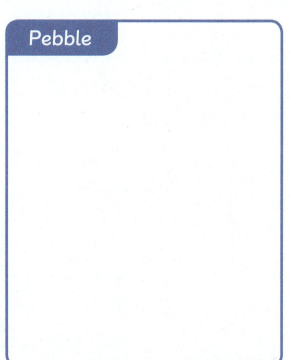

Pebble

Is this activity a good model for landscapes changing quickly or slowly? Why?

_____

_____

Activity 15
Evaluate Like a Scientist

Quick Code:
ca2521s

Water, Water, Water

Water erosion can change the land in different ways.
Look at the pictures.

Arch Along Lava Shore

Waterfalls

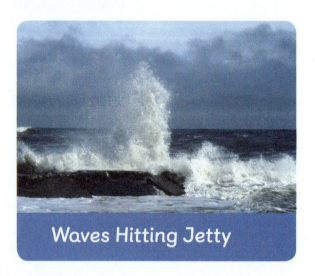

Waves Hitting Jetty

CCC Cause and Effect

© Discovery Education | www.discoveryeducation.com • Image: Images: (a) Paul Fuqua, (b) Paul Fuqua, (c) Paul Fuqua, Matteo Colombo / DigitalVision / Getty Images

Pick one of the pictures and **write** a story describing how water changed the landform over time.

© Discovery Education | www.discoveryeducation.com • Image: Matteo Colombo / DigitalVision / Getty Images

How Does a Land's Shape Change Quickly?

Activity 16
Analyze Like a Scientist

Quick Code: ca2523s

Landslides, Earthquakes, and Volcanoes

Read the text, **watch** the videos, and **look** at the picture.

Read Together

Landslides, Earthquakes, and Volcanoes

Landslides

Not all changes in a landscape happen slowly. Some changes happen quickly.

A landslide can move earth quickly. Gravity causes rocks, dirt, and boulders to speed down a hill.

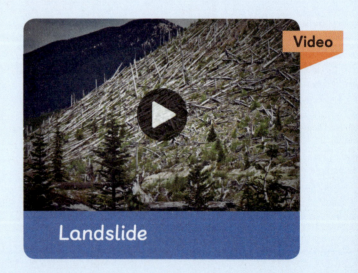

Video

Landslide

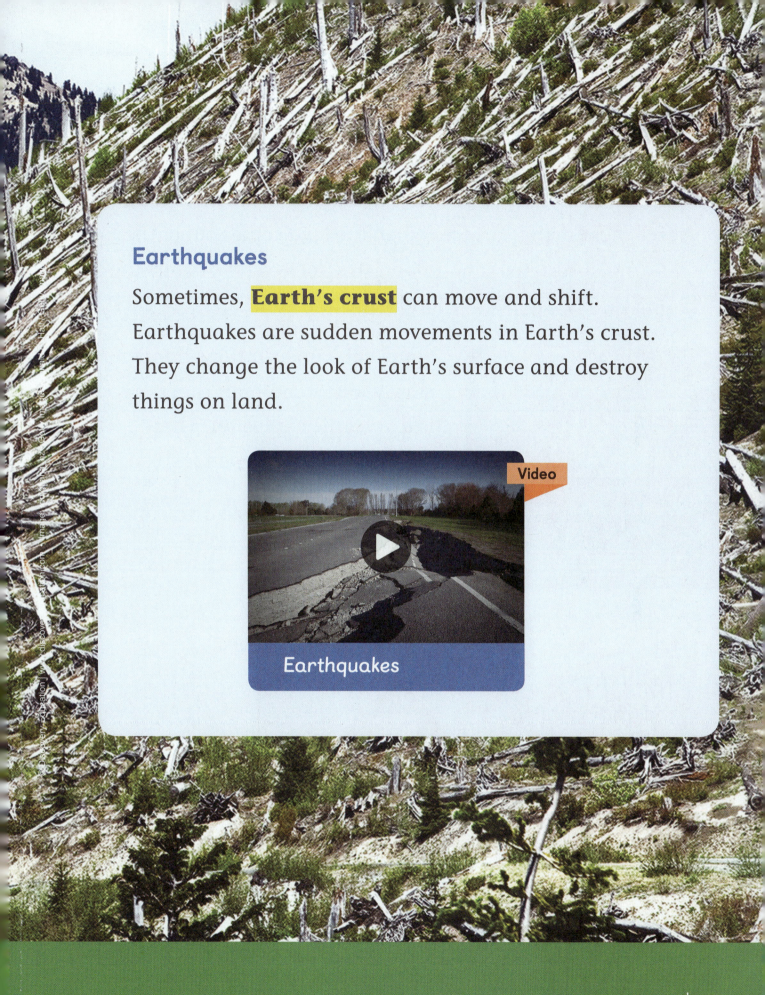

Earthquakes

Sometimes, **Earth's crust** can move and shift. Earthquakes are sudden movements in Earth's crust. They change the look of Earth's surface and destroy things on land.

Video

Earthquakes

Volcanoes

A volcano is an opening in Earth's crust. Liquid rock called magma rises in the volcano. The magma is called lava when it reaches Earth's surface. When the lava comes out of the volcano, it is called an eruption. The melted rock covers the landscape. It shapes the landscape as it cools into new rock.

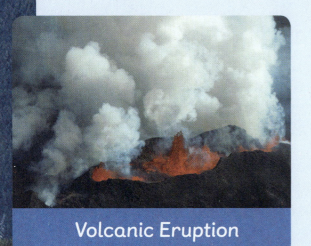

Volcanic Eruption

Some volcano eruptions happen slowly. But sometimes there is a volcanic explosion. These explosions can happen fast!

Go online to **explore** a volcano.

Interactive

Volcano

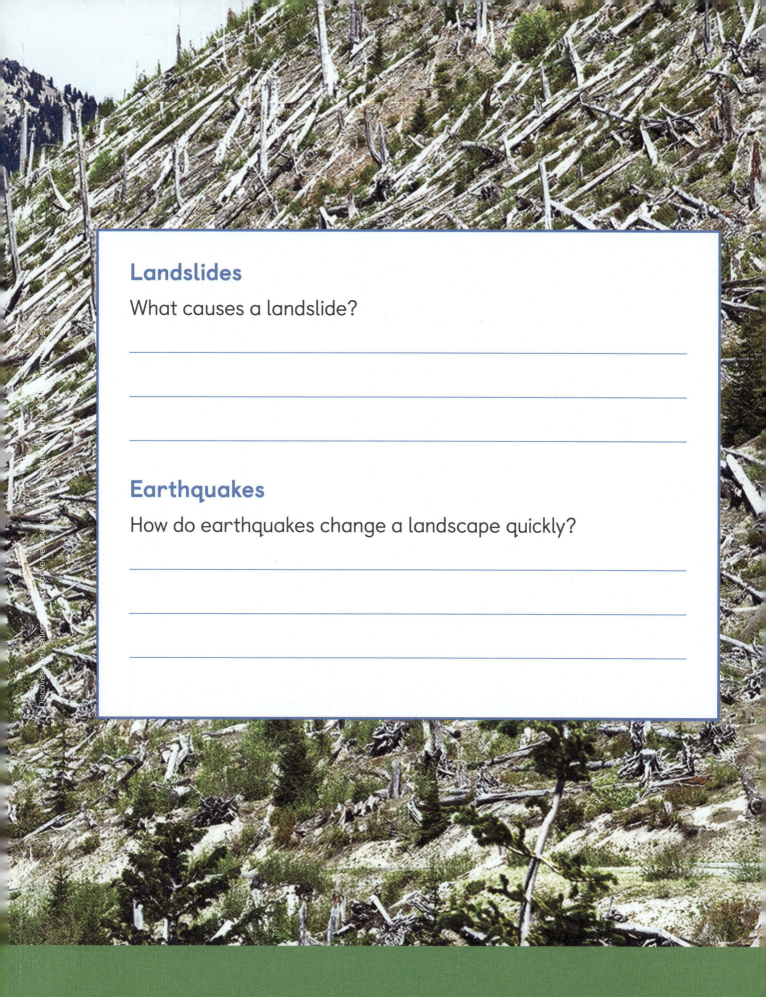

Landslides

What causes a landslide?

_____

_____

_____

Earthquakes

How do earthquakes change a landscape quickly?

_____

_____

_____

Volcanoes

How can a volcano change the landscape quickly?

_____

_____

_____

How does a volcano change the landscape? **Write** or **draw** your answer.

CCC Stability and Change

Activity 17
Evaluate Like a Scientist

Quick or Slow?

Look at each picture.

Draw a circle around the pictures that show slow changes.

Draw a square around the pictures that show quick changes.

Slow-Flowing River

Flash Flood

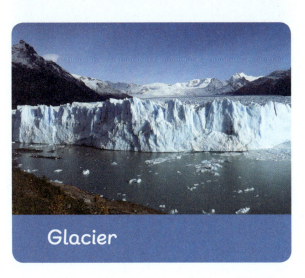

Glacier

Landslide

Activity 18
Record Evidence Like a Scientist

Quick Code:
ca2528s

House Mystery

Now that you have learned about how landscapes change, look again at the Let's Investigate a House Mystery photograph. You first saw this picture in Wonder.

Let's Investigate a House Mystery

 Talk Together

How can you describe the house mystery now? How is your explanation different from before?

| SEP | Constructing Explanations and Designing Solutions | CCC | Stability and Change |

Look at the Can You Explain? question. You first read this question at the beginning of the lesson.

Can You Explain?

How do landscapes change quickly or slowly over time?

Now, you will use your new ideas about Let's Investigate a House Mystery to answer a question.

1. **Choose** a question. You can use the Can You Explain? question, or one of your own. You can also use one of the questions that you wrote at the beginning of the lesson.

Your Question

2. Then, **use** the sentence starters on the next page to answer the question.

Landscapes can change slowing because of

_____

_____

_____

Landscapes can change quickly because of

_____

_____

_____

The evidence I collected shows

_____

_____

_____

_____

© Discovery Education | www.discoveryeducation.com • Image: Matteo Colombo / DigitalVision / Getty Images

STEM in Action

Activity 19
Analyze Like a Scientist

Protecting Land and Water

Read the text. Then, **answer** the questions.

Read Together

Protecting Land and Water

Some people work to save water and land. A park ranger might save a seashore from erosion. A ranger might work to keep water from washing trails away. Water that freezes and thaws can wash away land. A ranger might plant grasses to keep land from washing away.

SEP **Constructing Explanations and Designing Solutions**

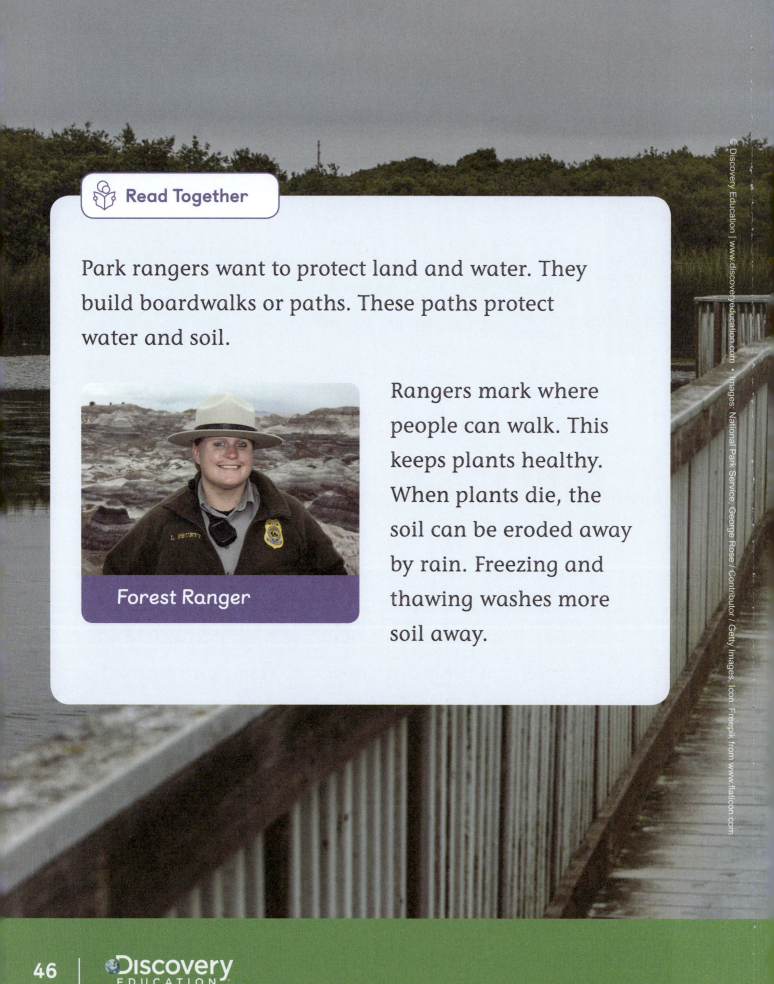

Read Together

Park rangers want to protect land and water. They build boardwalks or paths. These paths protect water and soil.

Forest Ranger

Rangers mark where people can walk. This keeps plants healthy. When plants die, the soil can be eroded away by rain. Freezing and thawing washes more soil away.

Discovery EDUCATION

What are other ways a ranger can keep water from washing away soil? **Write** or **draw** your answer.

Holding Soil in Place

Soil Washed Away by Water

How can farmers prevent soil from being washed away?

_____

_____

_____

_____

_____

Discovery
EDUCATION

Activity 20
Evaluate Like a Scientist

Quick Code:
ca2532s

Review: The Changing Landscape

Think about what you have read and seen. What did you learn?

Draw what you have learned. Then, tell someone else about what you learned.

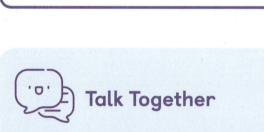

Talk Together

Think about what you saw in Get Started. Use your new ideas to discuss why landscapes change.

Landscape Solutions

Student Objectives

By the end of this lesson:

- [ ] I can design and test ways to slow or stop erosion caused by wind and water.

- [ ] I can make a model to show how the shape of an object helps solve problems caused from changes to the landscape.

Key Vocabulary

- [ ] barrier
- [ ] channel
- [ ] engineering
- [ ] habitat

Quick Code: ca2533s

Activity 1
Can You Explain?

How can we design solutions to solve problems caused by landscape changes?

_____

_____

_____

_____

_____

Quick Code:
ca2534s

Discovery
EDUCATION

Activity 2
Ask Questions Like a Scientist

Quick Code: ca2535s

A Breaking Dam

Watch the video. **Answer** the questions.

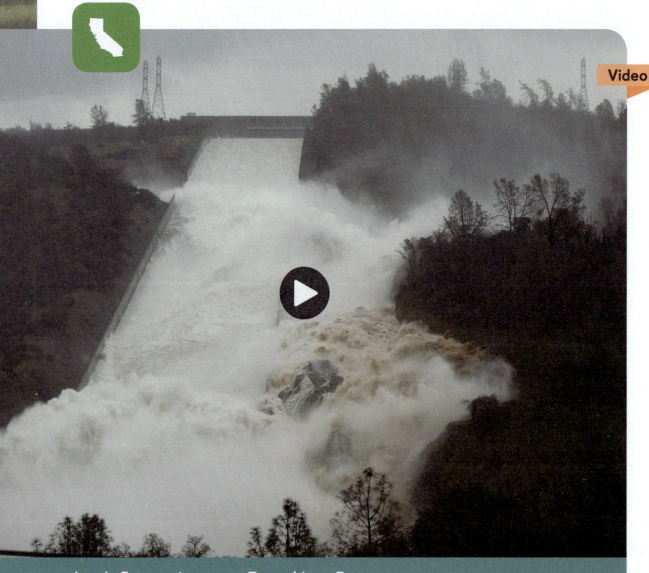

Let's Investigate a Breaking Dam

What were some problems with the Oroville Dam?

_____

_____

_____

What are some other ways landscape changes could cause problems? What questions do you have that would help you solve problems caused by landscape changes? **Draw** or **write** your ideas in the chart.

Problems from Landscape Changes	Questions About the Problems

Solutions

Beach Erosion

Water can wash away rocks and soil. This process is called erosion. Erosion can damage buildings and structures. **Engineers** must think about how the landscape can change when they make their designs.

Engineers are careful about the materials they choose for their designs. They also consider the type of weather in the area where they will place their designs.

Beach Erosion

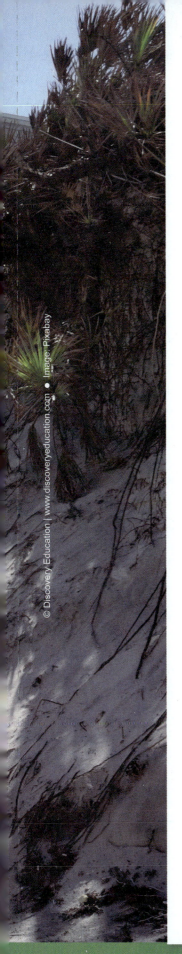

© Discovery Education | www.discoveryeducation.com • Image: Pixabay

Activity 3
Analyze Like a Scientist

Quick Code:
ca2537s

Beach Erosion

Look at the picture.

Draw your ideas for a design that would stop beach erosion.

Activity 4
Evaluate Like a Scientist

Quick Code:
ca2536s

What Do You Already Know About Landscape Solutions?

Weathering or Erosion?

Look at each picture. Which pictures show mostly weathering? Which pictures show mostly erosion? **Circle** the images that mostly show erosion.

Beach Coast

Large Rocks

Smooth Boulder

Sand Dunes

Pot Hole

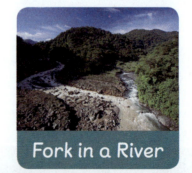

Fork in a River

Fill out the chart. **Tell** what you see in the pictures.
Write what you think about the pictures. What
do you wonder?

I see . . .

I think . . .

I wonder . . .

How Did It Happen?

Look at the pictures. **Tell** whether the shapes in each image were caused by wind, rivers, waves, or ice. Some shapes may have been caused by more than one. **Write** the word or words below the picture that describes the force (wind, river, waves, ice).

© Discovery Education | www.discoveryeducation.com • Image: Paola Moschitto-Assenmacher / Getty Images, wiratgasem / Moment / Getty Images

How Can We Fix Areas with Problems from Erosion?

Activity 5
Observe Like a Scientist

Quick Code:
ca2540s

Farming in the Dust Bowl

Look at the picture of a farm.

Dust Bowl

CCC Stability and Change

What do you think this farm looked like a month before this picture was taken? **Draw** your ideas.

What do you think this farm looked like a month after this picture was taken? **Draw** your ideas.

Farmer's Soil

Many years ago, farmers in the midwestern United States had a problem. For a long time, there was no rain. The top layers of soil got very dry. The wind began to blow a lot. The top layers of soil blew away in the wind. It was very dusty. When soil is moved from one location to another by wind or water, this is called erosion.

Soil Erosion

Farmers need soil for plants to grow. What can happen when the wind blows? The soil may blow away. What can happen if it rains? The soil may be washed away. Erosion can take away the farmer's soil and move it to a new location. Engineers look for ways to stop rocks and soil from moving away from places where they are needed.

Activity 6
Analyze Like a Scientist

Quick Code:
ca2541s

Farmer's Soil

Think about what you have read and the pictures you have seen.

Draw your ideas for a design that would stop erosion from changing a landscape.

How does the shape of your design help stop erosion?

_____

_____

_____

SEP Constructing Explanations and Designing Solutions
CCC Stability and Change

Activity 7
Observe Like a Scientist

Quick Code:
ca2542s

Erosion—Here Today, Gone Tomorrow

Use the Virtual Lab online to **test** some solutions for erosion.
Then, **talk** about your solutions.

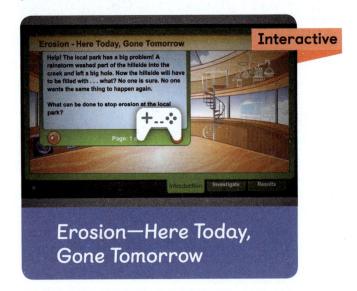

Erosion—Here Today,
Gone Tomorrow

💬 Talk Together

Talk together about your test solutions for erosion. What soil treatments did you test? How much water did you use? How steep was the slope for each?

SEP   Planning and Carrying Out Investigations

What happened in each of your tests? **Write** or **draw** what you saw in the chart below.

Test Number	How Much Soil Eroded?
1	
2	
3	

Which test ended with the least erosion?

_____

_____

Quick Code:
ca2545s

Slow or Stop Erosion

Read the text. As you read, **think** about the ways to stop erosion.

 Read Together

Slow or Stop Erosion

There are many ways to slow or stop erosion.

1. **Make a barrier to stop materials from moving.** Trees can block the wind from blowing away too much soil. Walls or fences can slow or prevent the movement of soil.

2. **Cover the land.** The roots of plants help keep soil in place. Wood chips or pine needles can cover the ground to protect soil.

3. **Direct the flow.** Rainwater can be directed from a rooftop, to a downspout, and into a rain garden. This prevents rainwater from eroding the soil near a building. Moist soil helps plants grow and is not easily blown by the wind. **Channels** can also be built to direct mudflows and falling rocks in a path away from buildings. A dam can control where water flows to prevent erosion.

SEP Engaging in Argument from Evidence

Activity 9
Investigate Like a Scientist

Quick Code:
ca2543s

Hands-On Investigation: Comparing Erosion Solutions

In this activity, you will look at different ways to prevent wind and water erosion.

Make a Prediction

You will make a landform out of soil. **Write** or **draw** your predictions.

What do you predict will prevent erosion by wind?

What do you predict will prevent erosion by water?

SEP Developing and Using Models

What materials do you need? (per group)

- Aluminum foil pan, 13 in. × 9 in. × 2 in.
- Soil, potting
- Watering can
- Water
- Disposable gloves (per student)
- Safety goggles (per student)

- Fan
- Artificial flowers or foliage (preferably grass, shrubs, and trees)
- Large bin, with lid
- Paper towels

HANDS-ON INVESTIGATION

What Will You Do?

Build your landform using the materials from your teacher.

Use the fan to blow wind over your landform. What happened?

Use the watering can to pour water over your landform.
What happened?

Place some trees, grass, and shrubs on your landform.
Then, let the fan blow over the plants and the landform.
What happened?

Next, **use** the watering can to pour water over the plants and the
landform. What happened?

Think About the Activity

Evaluate your solutions in the chart.

Solution	How Did It Work?
Plants	
Channels	
Plants and Channels	

Discovery EDUCATION

What impact did the shape of each solution have on preventing erosion by wind and water?

_____

_____

_____

_____

_____

What other solutions could you test to slow down or stop erosion?

_____

_____

_____

_____

Activity 10
Evaluate Like a Scientist

Quick Code:
ca2544s

Erosion Solutions

Look at the pictures. Which erosion solution do you think will work best?

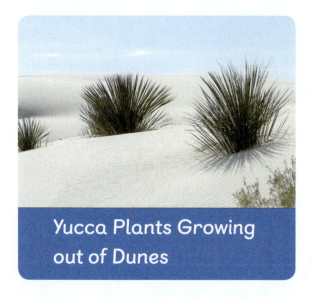

Yucca Plants Growing out of Dunes

Dam in Arizona

 **Talk Together**

Now talk together. What are the strengths of each solution? What are the weaknesses of each solution? Which do you think will work best?

SEP   Constructing Explanations and Designing Solutions

How Can We Fix Areas with Problems from Earthquakes?

 **Activity 11**
Analyze Like a Scientist

Quick Code:
ca2546s

Change from Earthquakes

Read the text and **look** at the pictures. As you read, **underline** the key idea.

 Read Together

Change from Earthquakes

Earthquakes change a landscape. They cause the ground to shake. Buildings and other structures can be damaged. People can be hurt.

Earthquake Damage

 CCC   **Structure and Function**

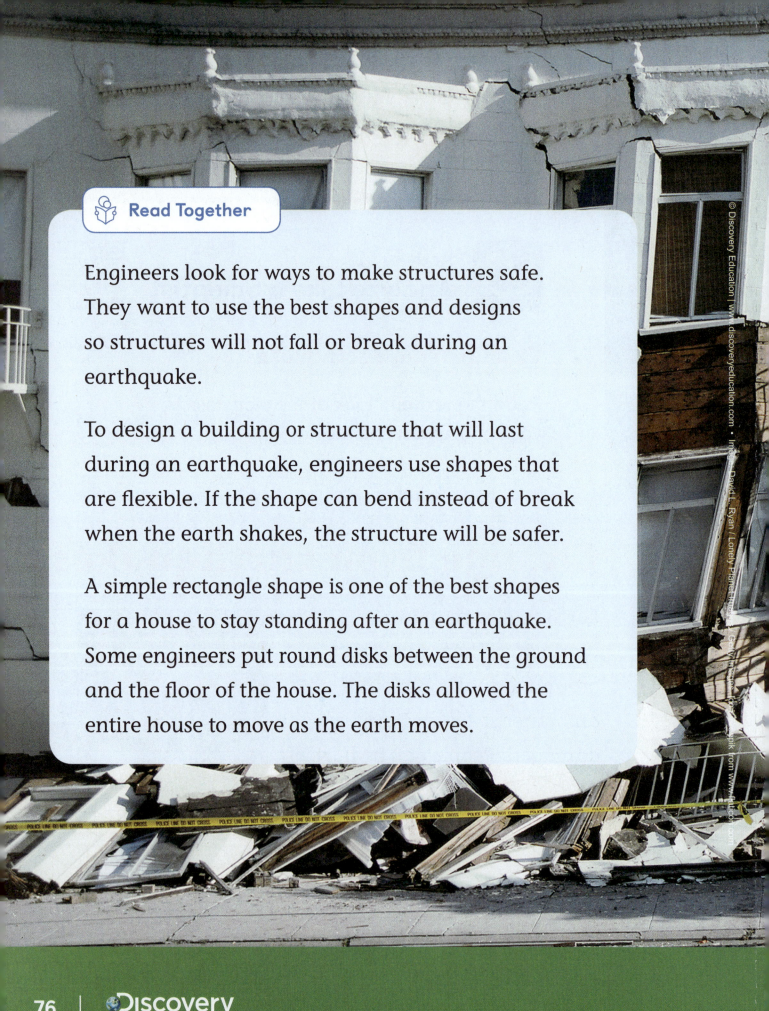

Read Together

Engineers look for ways to make structures safe. They want to use the best shapes and designs so structures will not fall or break during an earthquake.

To design a building or structure that will last during an earthquake, engineers use shapes that are flexible. If the shape can bend instead of break when the earth shakes, the structure will be safer.

A simple rectangle shape is one of the best shapes for a house to stay standing after an earthquake. Some engineers put round disks between the ground and the floor of the house. The disks allowed the entire house to move as the earth moves.

Discovery EDUCATION

Now, read the text again. **Circle** any words that tell about a surface or a shape changing.

Think about what you have learned. **Draw** a picture of a building that you think will last during an earthquake.

Activity 12
Design Solutions Like a Scientist

Quick Code:
ca2549s

Hands-On Engineering: Suspension Bridge

In this activity, you will use string, rubber bands, glue, wooden boards, and craft sticks to build a suspension bridge. **Think** about what you know about how engineers design buildings for earthquakes.

You will then test what happens when you remove parts of your design.

Ask About the Problem

You have to create a bridge that will hold a toy truck or a marble. **Write** down questions you have about the design challenge.

© Discovery Education | www.discoveryeducation.com

SEP Developing and Using Models   **SEP** Structure and Function

Discovery EDUCATION

What materials do you need? (per group)

- Wood blocks, 2
- Wooden boards
- Rubber bands
- String
- Craft sticks
- Glue
- Toy trucks

What Will You Do?

Look at the materials from your teacher. **Think** about what you have learned about shapes in designs. **Draw** one or two ideas for your bridge.

Build your bridge. **Roll** your toy truck or a marble across the bridge. It should stay on the bridge all the way across.
If it does not, **rebuild** your bridge and test it again.

Each time you test your bridge, **write** what you see in the chart below. **Draw** each change you make to your bridge design.

Test Number	Bridge Design	Result
1		
2		
3		

Discovery EDUCATION

Look at your bridge again. What pieces are not glued together? **Take out** one of those pieces. Then, **roll** your truck or a marble across the bridge again.

Draw or **write** what happened.

Think About the Activity

Did the bridge still work when you removed part of it?
Why or why not?

_____

_____

_____

_____

How did the materials work together to hold up the truck or
marble? **Draw** or **write** how each material helped the bridge to
work well.

Discovery EDUCATION

Material

Effect

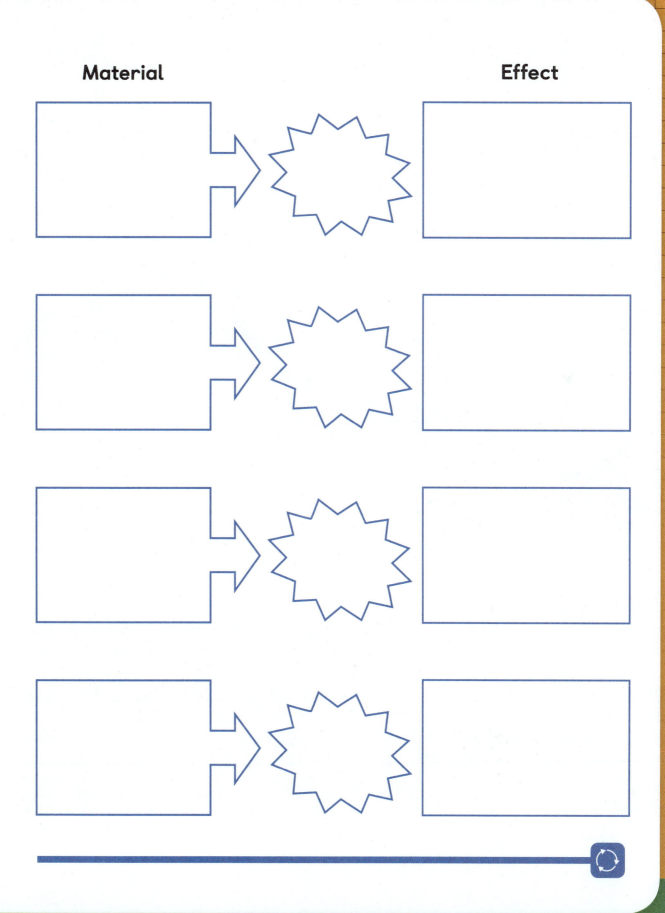

Activity 13
Evaluate Like a Scientist

Quick Code:
ca2547s

Under Construction

Look at the picture of construction workers building the house.

Constructing a House

SEP Constructing Explanations and Designing Solutions

What could you do to make the house better so it won't get damaged by landscape changes?

Draw a new picture of the house showing one thing you would change to stop damage from erosion. Then, show one thing you would change to stop damage from an earthquake.

Activity 14
Record Evidence Like a Scientist

Quick Code:
ca2551s

Breaking Dam

Now that you have learned more about landscape solutions, look again at the Let's Investigate a Breaking Dam video. You first saw this in Wonder.

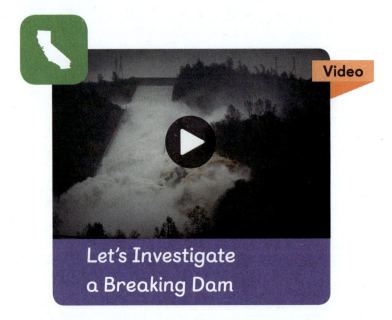

Video

Let's Investigate
a Breaking Dam

 Talk Together

How can you describe the solutions in the Breaking Dam video now? How is your explanation different from before?

SEP Constructing Explanations and Designing Solutions

© Discovery Education | www.discoveryeducation.com • Image: AP / Shutterstock, wiratgasem / Moment / Getty Images, Icon: Freepik from www.flaticon.com

Look at the Can You Explain? question. You first read this question at the beginning of the lesson.

Can You Explain?

How can we design solutions to solve problems caused by landscape changes?

Now, you will use your new ideas about Let's Investigate a Breaking Dam to answer a question.

1. **Choose** a question. You can use the Can You Explain? question, or one of your own. You can also use one of the questions that you wrote at the beginning of the lesson.

Your Question

2. Then, **use** the sentence starters on the next page to answer the question.

A design solution to wind and water erosion would be

_____

_____

_____

_____

A design solution to earthquake damage would be

_____

_____

The evidence I collected shows

_____

_____

_____

_____

STEM in Action

Activity 15
Analyze Like a Scientist

Knowing About Weathering and Erosion for a Job

 Read Together

Knowing About Weathering and Erosion for a Job

Weathering wears down rocks. It breaks cliffs. Erosion carries away the small rocks. It moves sand. Human activity can cause erosion. Chopping down trees leaves the soil bare. Wind and rain can move soil away. People can help stop erosion. They can think of ways to stop soil from moving. It is an important job to stop erosion. Physical geography engineers work on problems of erosion.

SEP Obtaining, Evaluating, and Communicating Information

Read Together

Each kind of erosion has a different cause. The engineer will need to design a new way to stop each kind of erosion. People build walls to stop soil from sliding.

People plant more trees and plants. The roots hold the soil in place. People plant trees to block the wind.

Beach Erosion

Discovery
EDUCATION

What other ways can people help stop erosion?

Saving the Beach

A 50-meter stretch along the beach is a favorite spot for sea turtles to lay their eggs. The wind and waves are washing the beach away.

Your class wants to save the ==habitat== for turtles.

Beach Fence

If you can spend only $50, what things would you buy to install on the dune to save a turtle habitat and stop sand from blowing away?

Materials	
$10—2 meters of fencing	$4—a bush to plant
$3—10 dune grass plants	$2—a sign to warn people not to walk on the dune

List your materials and **show** your costs below.

Write or **draw** how these items will help stop beach erosion.

Activity 16
Evaluate Like a Scientist

Quick Code:
ca2557s

Review: Landscape Solutions

Think about what you have read and seen. What did you learn?

Draw what you have learned. Then, **tell** someone else about what you learned.

Talk Together

Think about what you saw in Get Started. Use your new ideas for ways to solve problems caused by landscape changes.

Design Solutions Like a Scientist

Quick Code:
ca2552s

Hands-On Engineering:
Landscape Changes

In this activity, you will design a solution to reduce or prevent landscape erosion.

Crumbling Cliff

 SEP Analyzing and Interpreting Data  **CCC** Structure and Function

What materials do you need? (per group)

- Paper
- Colored pencils
- Metric ruler
- Books and/or magazines with landscape designs
- Access to Internet to gather images (optional)
- Various materials based on student designs (stone, plants, mulch, etc.)

HANDS-ON ENGINEERING

Ask Questions About the Problem

Locate an area that has a problem with erosion. What questions do you have about the area before you can design a solution?

_____

_____

_____

_____

_____

Unit Project

What Will You Do?

Look at the slope of the landscape you selected.

How can you slow down erosion in your landscape?

_____

_____

_____

Test your design. **Draw** a picture or **use** words to show how you tested your design.

Discovery EDUCATION

Think About the Activity

Write or **draw** your answers to the questions in the chart.
How well did your design prevent or reduce erosion?
How could you improve your design?

What Worked?	What Didn't Work?

What Could Work Better?

Grade 2 Resources

- Bubble Map
- Safety in the Science Classroom
- Vocabulary Flash Cards
- Glossary
- Index

Name _____

Bubble Map

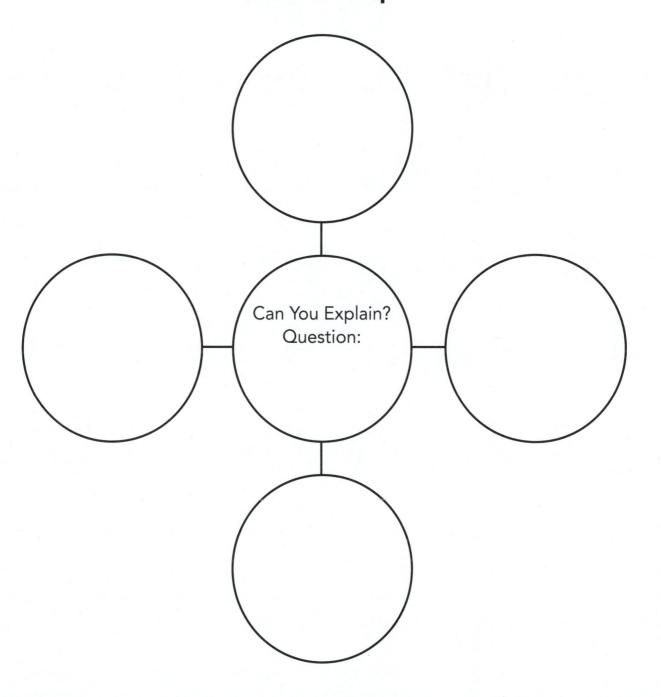

Can You Explain?
Question:

Following common safety practices is the first rule of any laboratory or field scientific investigation.

Dress for Safety

One of the most important steps in a safe investigation is dressing appropriately.

- Splash goggles need to be kept on during the entire investigation.

- Use gloves to protect your hands when handling chemicals or organisms.

- Tie back long hair to prevent it from coming in contact with chemicals or a heat source.

- Wear proper clothing and clothing protection. Roll up long sleeves, and if they are available, wear a lab coat or apron over your clothes. Always wear closed-toe shoes. During field investigations, wear long pants and long sleeves.

Safety Goggles

Be Prepared for Accidents

Even if you are practicing safe behavior during an investigation, accidents can happen. Learn the emergency equipment location in your classroom and how to use it.

- The eye and face wash station can help if a harmful substance or foreign object gets into your eyes or onto your face.

- Fire blankets and fire extinguishers can be used to smother and put out fires in the laboratory. Talk to your teacher about fire safety in the lab. He or she may not want you to directly handle the fire blanket and fire extinguisher. However, you should still know where these items are in case the teacher asks you to retrieve them.

Most importantly, when an accident occurs, immediately alert your teacher and classmates. Do not try to keep the accident a secret or respond to it by yourself. Your teacher and classmates can help you.

Practice Safe Behavior

There are many ways to stay safe during a scientific investigation. You should always use safe and appropriate behavior before, during, and after your investigation.

- Read all of the steps of the procedure before beginning your investigation. Make sure you understand all the steps. Ask your teacher for help if you do not understand any part of the procedure.

- Gather all your materials and keep your workstation neat and organized. Label any chemicals you are using.

- During the investigation, be sure to follow the steps of the procedure exactly. Use only directions and materials that have been approved by your teacher.

- Eating and drinking are not allowed during an investigation. If asked to observe the odor of a substance, do so using the correct procedure known as wafting, in which you cup your hand over the container holding the substance and gently wave enough air toward your face to make sense of the smell.

- When performing investigations, stay focused on the steps of the procedure and your behavior during the investigation. During investigations, there are many materials and equipment that can cause injuries.

- Treat animals and plants with respect during an investigation.

- After the investigation is over, appropriately dispose of any chemicals or other materials that you have used. Ask your teacher if you are unsure of how to dispose of anything.

- Make sure that you have returned any extra materials and pieces of equipment to the correct storage space.

- Leave your workstation clean and neat. Wash your hands thoroughly.

barrier

Image: Nutthika / Shutterstock.com

something that is used to stop or block materials from moving

channel

Image: Paul Fuqua

a path that is dug and used for drainage or protection against things like water, mud, or rocks

Earth's crust

Image: Paul Fuqua

the top layer of Earth that is the thinnest and the most important because it is where we live

earthquake

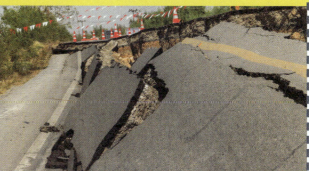

Image: Naypong Studio

a sudden shaking of the ground caused by the movement of rock underground

engineering

using math and science to design and build machines, structures, and other devices

erosion

when soil is moved from one location to another by wind or water

habitat

the place where a plant or animal lives

landform

a feature of Earth that has been formed by nature, such as a hill or a valley

landscape

Image: Discovery Communications

the view of a land's surface

material

Image: Peter Menzel Photography

things that can be used to build or create something

weathering

Image: Paul Fuqua

the breakdown of rocks into smaller pieces called sediment

English ———— **A** ———— **Español**

absorb
to take in or soak up

absorber
tomar o captar

absorption
how much something can take in and hold

absorción
cuanto algo puede tomar y retener

adjust
to fix or change something

ajustar
arreglar o cambiar algo

analyze
to closely examine something and then explain it

analizar
examinar con atención algo y luego explicarlo

———— **B** ————

barrier
something that is used to stop or block materials from moving

barrera
algo que se usa para evitar o bloquear el movimiento de materiales

biodiversity

the many different types of life that live together in an environment

biodiversidad

muchos y diferentes tipos de vida que conviven en un medio ambiente

—————— **C** ——————

canyon

a deep valley that has very steep sides

cañón

valle profundo que tiene laderas muy pronunciadas

channel

a path that is dug and used for drainage or protection against things like water, mud, or rocks

canal

vía cavada que se usa como desagüe o protección contra cosas como el agua, el lodo o las rocas

characteristic

a special quality that something may have

característica

cualidad especial que tiene algo

—————— **D** ——————

dissolve

to mix something with a liquid, such as water, so that it can't be seen anymore

disolver

mezclar algo con un líquido, como el agua, de manera que no se pueda ver más

drought

when there is no rain for a long period

sequía

cuando no llueve durante un período prolongado

———————— E ————————

Earth's crust

the top layer of Earth that is the thinnest and the most important because it is where we live

corteza de la Tierra

capa superior de la Tierra que es la más delgada y la más importante porque allí es donde vivimos

earthquake

a sudden shaking of the ground caused by the movement of rock underground

terremoto

repentina sacudida de la tierra causada por el movimiento de roca subterránea

elevation

the height of an area of land above sea level

elevación

altura de un área de tierra por encima del nivel del mar

engineer

a person who designs something that may be helpful to solve a problem

ingeniero

persona que diseña algo que puede ser útil para resolver un problema

engineering

using math and science to design and build machines, structures, and other devices

ingenería

usar las matemáticas y las ciencias para diseñar y construir máquinas, estructuras y otros dispositivos

---

environment

all the living and nonliving things that surround an organism

medio ambiente

todos los seres vivos y objetos sin vida que rodean a un organismo

---

erosion

when soil is moved from one location to another by wind or water

erosión

cuando el viento o el agua transporta suelo de un lugar a otro

---

estimate

to make a careful guess

estimar

hacer una suposición consciente

——— **F** ———

feature

a thing that describes what something looks like; part of something

rasgo

cosa que describe cómo se ve algo; parte de algo

flexibility

the ability to bend without breaking

flexibilidad

capacidad de doblarse sin romperse

---

fresh water

water that is not salty, such as that found in streams and lakes

agua dulce

agua que no es salada, como la que se encuentra en arroyos y lagos

—— G ——

gemstone

a colorful stone found in nature that can be used for jewelry

piedra preciosa

piedra colorida que se encuentra en la naturaleza y se puede usar para hacer joyas

—— H ——

habitat

the place where a plant or animal lives

hábitat

lugar donde vive una planta o un animal

---

hardness

a measure of how difficult it is to scratch a mineral: Diamonds are the hardest mineral. They have a hardness scale rating of 10.

dureza

medida de cuán difícil es rayar un material: los diamantes son los minerales más duros. Su clasificación en la escala de dureza es 10.

L

landfill
a place where trash is buried

vertedero
lugar donde se entierra la basura

landform
a feature of Earth that has been formed by nature, such as a hill or a valley

accidente geográfico
característica de la Tierra formada por la naturaleza, como una colina o un valle

landscape
the view of a land's surface

paisaje
vista de la superficie de un terreno

location
a place where something is

ubicación
lugar donde se encuentra algo

M

map
a flat picture or drawing of a place that is made to show things, such as streets or towns, in an area

mapa
imagen o dibujo plano de un lugar que se hace para mostrar cosas, como las calles o las ciudades, de un área

material

things that can be used to build or create something

material

cosas que se pueden usar para construir o crear algo

---

matter

the things around you that take up space like solids, liquids, and gases

materia

cosas que nos rodean y ocupan espacio, como los sólidos, los líquidos y los gases

---

mixture

a combination of different things, but you can pick out each different one

mezcla

combinación de diferentes cosas, pero se puede identificar cada una

---

model

a human-made version created to show the parts of something else, either big or small

modelo

versión creada por el hombre para mostrar las partes de algo más, ya sea grande o pequeño

---

mountain

a very tall area of land that is higher than a hill and has steep sides

montaña

área de tierra muy alta que es más alta que una colina y tiene laderas pronunciadas

N

naturalist
someone who studies nature, especially plants and animals

naturalista
alguien que estudia la naturaleza, especialmente las plantas y los animales

nutrient
something in food that helps people, animals, and plants live and grow

nutriente
algo en los alimentos que ayuda a las personas, los animales y las plantas a vivir y crecer

O

observe
to watch closely

observar
mirar atentamente

ocean
a large body of salt water

océano
gran cuerpo de agua salada

organism
a living thing

organismo
ser vivo

P

plain
a large flat area of land without trees

llanura
gran área de tierra llana sin árboles

plateau

a large, flat area of land that is higher than the other land around it

meseta

gran área de tierra llana que está a más altura que el terreno que la rodea

pollen

the yellow powder found inside a flower

polen

polvo amarillo que se encuentra dentro de una flor

pollination

moving or carrying pollen from a plant to make the seeds grow

polinización

transferencia o transporte de polen de una planta para hacer que crezcan las semillas

preserve

to protect or keep something safe

preservar

proteger o mantener algo a salvo

property

a characteristic of something

propiedad

característica de algo

— Q —

quadrilateral

a flat shape with four straight sides, such as a square or a parallelogram

cuadrilátero

figura plana con cuatro lados rectos, como un cuadrado o un paralelogramo

R

recycle
to create new materials from something already used

reciclar
crear nuevos materiales a partir de algo usado

relief map
a type of map that shows how flat or steep the landforms are in an area

mapa de relieve
tipo de mapa que muestra si los accidentes geográficos son llanos o pronunciados en un área

resource
a material that can be used to solve problems

recurso
material que se puede usar para resolver problemas

restore
to put into use again

restablecer
volver a poner en servicio

reverse engineering
the process of learning about something by taking it apart to see how it works and what it is made of

ingeniería inversa
proceso de aprender acerca de algo, desarmándolo para ver cómo funciona y de qué está hecho

river
water flowing through a landscape, usually fed by smaller streams

río
agua que fluye a través de un área, por lo general alimentada por arroyos más pequeños

—————— **S** ——————

select
to choose or pick

seleccionar
elegir o escoger

shelter
a place that protects you from harm or bad weather

refugio
lugar para protegerse de peligros o el mal tiempo

slope
land that is slanted or angles downward

pendiente
tierra inclinada hacia abajo

soil
dirt that covers Earth, in which plants can grow and insects can live

suelo
tierra que cubre nuestro planeta en la que pueden crecer plantas y vivir insectos

solution
a combination of two things that are mixed so well that each one cannot be picked out

solución
combinación de dos cosas que se mezclan tan bien que no se puede identificar cada una

strategy
a plan that can solve a problem

estrategia
plan que puede resolver un problema

stream
a small flowing body of water that starts with a spring and ends at a river

arroyo
pequeño cuerpo de agua que fluye y nace en una vertiente y termina en un río

survive
to continue to live

sobrevivir
continuar viviendo

— T —

two-dimensional
drawings and sketches that are done on flat paper to show width and height

bidimensional
dibujos y bosquejos que se hacen en papel plano para mostrar el ancho y la altura

V

valley
the low place between two hills or mountains

valle
lugar bajo entre dos colinas o montañas

W

weathering
the breakdown of rocks into smaller pieces called sediment

meteorización
desintegración de rocas en trozos más pequeños llamados sedimento

Index

Discovery EDUCATION